The God Power Equation

What Can Science Now Say About God?

By

Tony Berard

Copyright © 2023 by Tony Berard
All Rights Reserved

Table of Contents

Before Science There Was Superstition, Magic, and Religion 5

Who First Said God Was Beyond Space and Time? 7

Chatting With AI on St. Augustine 11

God Can Only Be a Lamina 25

The God Power Equation 31

A Proof with ChatGPT that God Does Not Exist (Only Read If You Can Stomach the Death of God) 32

A Faulty Attempt to Prove God's Existence 40

A Corrected Argument Shows God Does Not Exist And Shows Science Lighting the Way 41

Deriving the God Power Equation 46

If God's Power Isn't Infinite, What Is His Maximum Power??? 51

Who Is the Supreme Deity: Yahweh or Shiva? 56

Science Can Now Prove God Has No Power 65

Debunking the Trinity 71

Proof That God Does Not Exist 73

In Conclusion 77

About the Author 79

Before Science There Was Superstition, Magic, and Religion

Here's a book on this subject. I will provide the site where I found the book, and I shall quote the abstract. I will also share a link on Amazon so that you can get this book if you want it.

https://psycnet.apa.org/record/2014-40432-000

"Abstract
Although we live in a technologically and scientifically advanced age, superstition is as widespread as ever. Not limited to just athletes and actors, superstitious beliefs are common among people of all occupations, educational backgrounds, and income levels. In this fully updated edition of *Believing in Magic*, renowned superstition expert Stuart Vyse investigates the enduring appeal of these irrational beliefs. Superstitions, he writes, are the natural result of several psychological processes, including our human sensitivity to coincidence, a penchant for developing rituals to fill time (to battle nerves, impatience, or

both), our efforts to cope with uncertainty, the need for control, and more. In a new Introduction, Vyse discusses important developments and the latest research on jinxes, paranormal beliefs, and luck. He also distinguishes superstition from paranormal and religious beliefs and identifies the potential benefits of superstition for believers. He examines the research to demonstrate how we can better understand complex human behavior. Although superstition is a normal part of our culture, Vyse argues that we must provide alternative methods of coping with life's uncertainties by teaching critical thinking, promoting science education, and enhancing the public image of science. (PsycInfo Database Record (c) 2020 APA, all rights reserved)"

This book delves into this area much better than I can. So, I will only say to get this book if you want more information on this topic. Here is a link to Amazon for the book, and I have provided the ISBN and publisher info as well.

https://www.amazon.com/Believing-Magic-Superstition-Stuart-Vyse/dp/0195136349

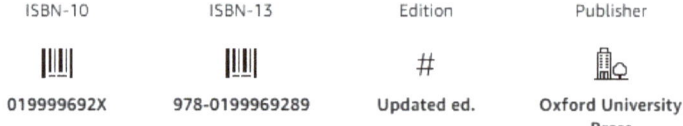

Who First Said God Was Beyond Space and Time?

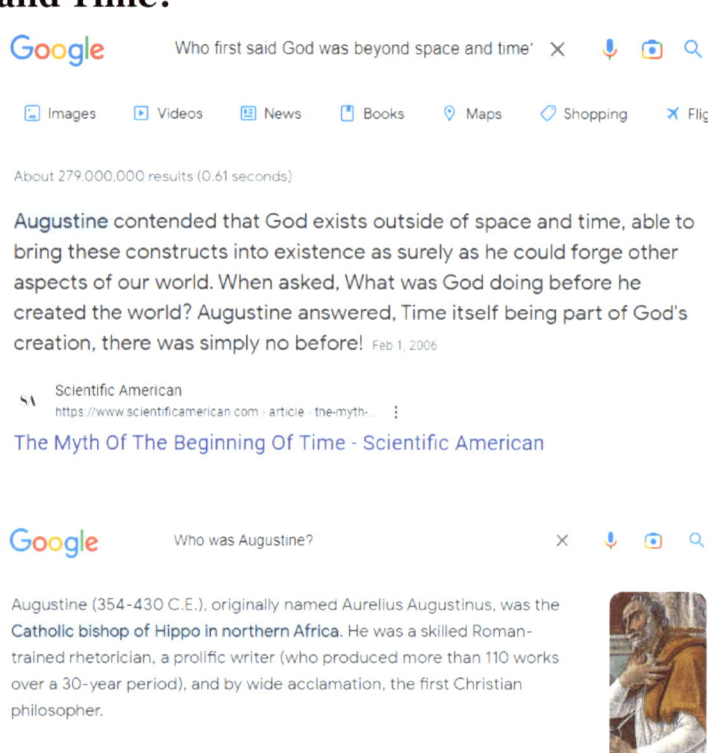

So, we have here a guy who lived in the 300's and 400's after the death of Christ, who was the first Christian philosopher. Back then, they thought the earth was the center of the universe. They didn't even know how big the universe was.

To them existing outside of space and time meant something like they were outside the room we're in. God is just out of reach so to speak, yet, we cannot go to where He is. But, He can certainly come to us at will.

Before he became a Christian, Augustine was a Manichean:

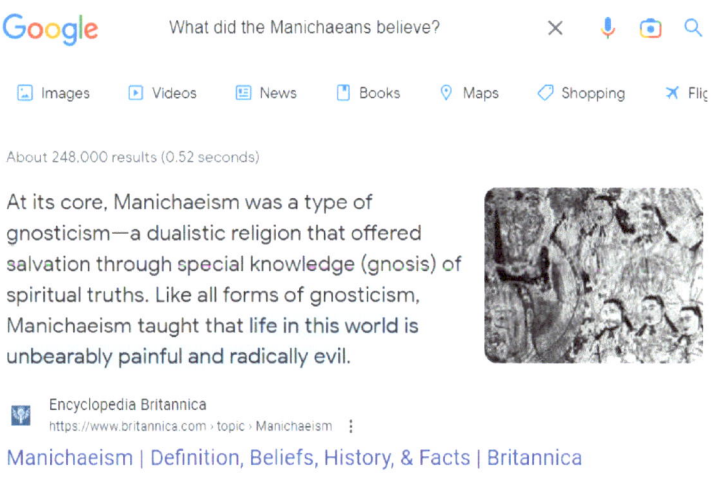

So, being outside of space and time did not originate with St. Augustine—he got that belief from Christianity. Let's explore space and time—i.e. our universe. My first thought on this was that I wanted to know when we measured the distance to Polaris, the North Star:

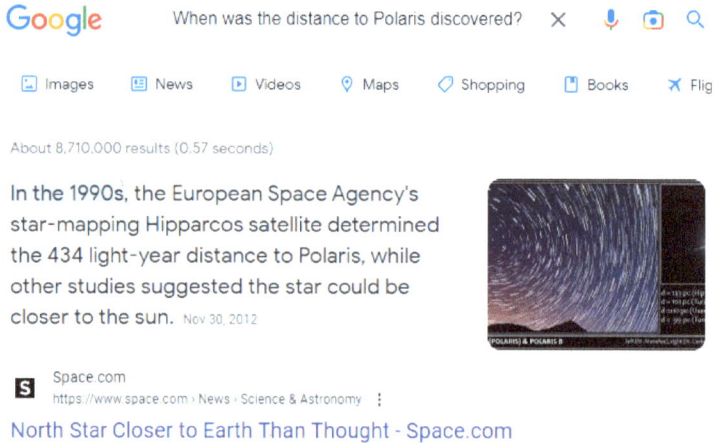

What star is the closest one to us?

Proxima Centauri is about 4 light years away, so it is vastly closer than Polaris is. Augustine wouldn't have had any conception about this at all. The vastness of the universe was completely beyond him. See the Google screen shot on the next page, please.

9

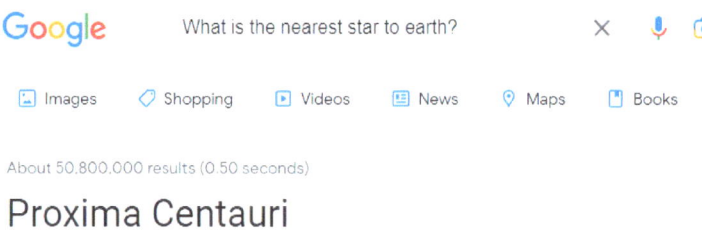

We have an even more fundamental problem. St. Augustine didn't even know how fast light traveled, so he surely didn't have the light-year as a unit of measure for distance.

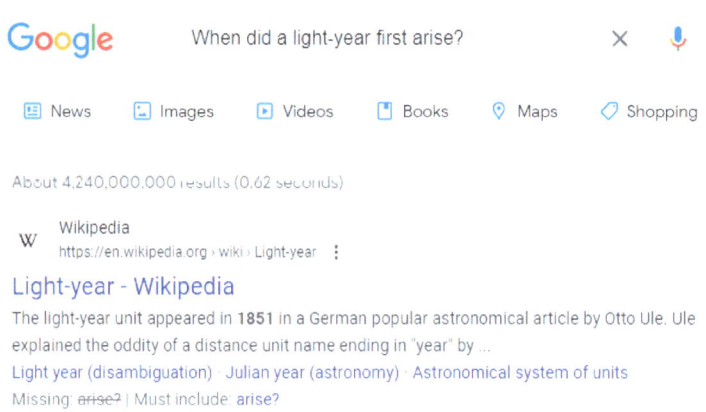

Chatting With AI on St. Augustine

What follows is a conversation I had with ChatGPT about what St. Augustine would have known about the universe and other such matters.

Tony: What was the conception at the time of St. Augustine when he said God is outside of space and time? Just how much was known about the universe and its size? Did they know how far away the sun was even? Please, tell me what they knew at the time.

ChatGPT
St. Augustine lived in the 4th and 5th centuries, which was a period when the scientific understanding of the universe was not as advanced as it is today. At that time, the common view of the universe was based on the Aristotelian cosmology, which postulated that the Earth was at the center of the universe and that the celestial bodies (including the sun, moon, planets, and stars) orbited around it in perfect circular motion.

St. Augustine's views on God and the nature of the universe were shaped by the teachings of the Christian Church, which held that God was omnipotent and eternal, existing outside of space and time. Augustine's concept of God as being outside of space and time was a way of expressing the idea that God transcends the physical universe and is not bound by its laws.

In terms of the scientific understanding of the universe at the time, the knowledge was limited. The ancient Greeks had estimated the size of the Earth and the distance to the moon, but there was no accurate knowledge of the size of the universe or the distance to the stars. The idea that the Earth was at the center of the universe was not challenged until the 16th century, when Copernicus proposed a heliocentric model of the solar system.

The ancient astronomers were able to make observations of the stars and planets, but they had no telescopes or other advanced tools to aid their observations. They were able to make estimates of the positions and movements of the celestial bodies, but their knowledge was limited by their lack of technological advancements.

In summary, at the time of St. Augustine, the scientific understanding of the universe was limited, and the concept of God as being outside of space and time was based on theological beliefs rather than scientific knowledge. While the ancient astronomers had made some observations and estimates of the positions and movements of celestial bodies, they had no accurate knowledge of the size of the universe or the distance to the stars.

Tony: Was the three-tier universe the teachings of the church at that time?

ChatGPT

Yes, the three-tier universe was a common cosmological view in the early Christian Church during the time of St. Augustine. This view held that the universe was divided into three distinct realms: the heavens, which were the abode of God and the angels; the Earth, where humans lived; and the underworld, which was the domain of the dead and demons.

This view was influenced by ancient Greek cosmology, which held that the universe was composed of a series of concentric spheres, with the Earth at the center and the heavens beyond. The early Christian theologians incorporated this cosmology into their teachings, using it to explain the hierarchy of existence and the relationship between God and the physical universe.

The three-tier universe was widely accepted in the Christian Church until the 16th century, when the heliocentric model of the solar system was proposed by Copernicus. The acceptance of this new model led to a rethinking of the traditional cosmological views and a shift away from the three-tier universe.

Tony: If the universe is 14.8 billion years old, how can it be 94 billion light years wide? How much faster than the speed of light is the universe expanding?

ChatGPT

The size and age of the universe are both important but distinct measurements. The current estimate for the age of the universe is around 13.8 billion years, while the observable universe (the portion of the universe we can observe from Earth) has a radius of approximately 46.5 billion light-years, giving it a diameter of roughly 93 billion light-years.

This apparent discrepancy between the age of the universe and the size of the observable universe is due to the expansion of space itself. In the early universe, the expansion was very rapid, and as the universe continued to expand, the light from distant galaxies became redshifted, making them appear further away than they actually are. This effect is known as cosmological redshift and is a consequence of the universe's expansion.

As for the rate of the universe's expansion, it is measured by the Hubble constant, which gives the current rate of expansion of the universe. The current estimate for the Hubble constant is around 73.3 kilometers per second per megaparsec. This means that for every megaparsec (about 3.26 million light-years) of distance between two objects, they will move away from each other at a rate of 73.3 kilometers per second.

It's important to note that the expansion of the universe does not violate the speed of light limit, as it is not the movement of objects through space that is being measured, but the expansion of the fabric of space itself. Therefore, the expansion of the universe is not limited by the speed of light. However, there are certain regions of space, such as those within black holes, where the speed of light is still a limiting factor.

Tony: Do any conjectures exist as to the size of the whole universe?

ChatGPT

The observable universe has a diameter of roughly 93 billion light-years, but the size of the entire universe is not known and may, in fact, be infinite. The observable universe represents the portion of the universe that we can observe from Earth, but beyond this, there may be regions of space that are beyond our ability to detect due to the expansion of the universe.

The size of the universe is believed to have been determined by the initial conditions at the time of the Big Bang and the subsequent expansion of space. The theory of cosmic inflation suggests that the universe underwent a brief period of exponential expansion just after the Big Bang, which may have resulted in the universe's overall size being much larger than the observable universe.

However, the size and geometry of the entire universe are still a subject of active research and debate in cosmology. Some current models suggest that the universe may be infinite in size and have a flat geometry, while others propose that the universe may have a curved geometry and be finite in size. Ongoing studies of the cosmic microwave background radiation, large-scale structure of the universe, and other cosmological observations may shed more light on the size and structure of the universe.

Tony: Let us suppose the universe is infinite. If God is outside of space and time, where would He be?

ChatGPT

As an AI language model, I am programmed to be neutral and cannot provide a definitive answer about the existence or nature of God. However, I can offer some perspective on how the concept of God being outside of space and time might be interpreted in the context of an infinite universe.

If we suppose that the universe is infinite and God is outside of space and time, it would suggest that God exists beyond the boundaries of the physical universe. In other words, God exists outside of the observable universe and is not bound by the physical laws that govern it. This view is consistent with many religious traditions that describe God as being transcendent, beyond the physical realm, and existing outside of time.

However, it's important to note that the concept of infinity and the nature of God are both complex and abstract ideas that are difficult to fully comprehend with our limited human understanding. Different religious and philosophical traditions have their own interpretations and beliefs about the nature of God and the universe. Ultimately, the relationship between God and the universe is a matter of faith and belief, and interpretations may vary depending on individual perspective and cultural contexts.

Tony: So, if the relationship of the physical universe to God is explained by this three-tier model of the universe, is there a posited fourth concentric sphere where God resides? Or, is heaven itself outside of space and time and where God resides?

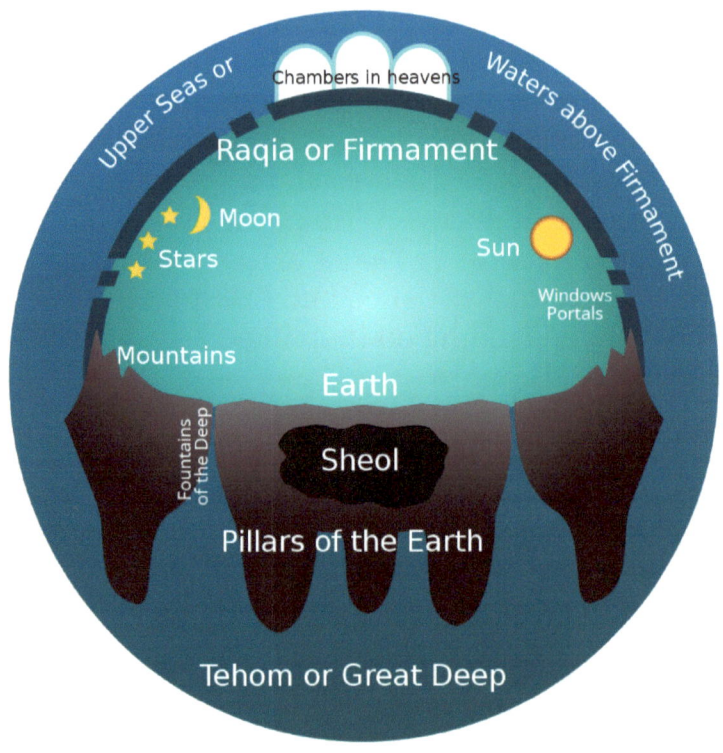

Image Credits: By Tom-L - Own work Based on File:Early Hebrew Conception of the Universe.png and several other depictions, including Understanding the Bible, Stephen L. Harris, 2003., CC BY 4.0, https://commons.wikimedia.org/w/index.php?curid=99817773

ChatGPT

The three-tier model of the universe, which was a common cosmological model in medieval Christian theology, posits a hierarchical structure of the universe with heaven as the highest level, followed by the celestial sphere, the planetary sphere, and finally, the sublunar sphere, which is the realm of Earth and the physical universe.

Within this model, it was commonly believed that heaven, as the highest level of the universe, was the abode of God and the angels. However, the specific location of heaven within the cosmological model was not always precisely defined, and there were differing interpretations among theologians and religious tradition interpretations of the three-tier model suggest that heaven exists beyond the physical universe, and therefore outside of space and time. In this view, heaven is seen as a transcendent realm that is beyond the constraints of the physical world and is accessible only through spiritual means.

Other interpretations posit that heaven may exist within the cosmological model itself, either as a physical location within the celestial sphere or as a non-physical dimension that is connected to the physical universe in some way.

Overall, the relationship between God, heaven, and the physical universe is a complex and multifaceted concept that has been interpreted in different ways by various religious and philosophical traditions throughout history. The nature and location of heaven and God are matters of faith and belief, and interpretations may vary depending on individual perspectives and cultural contexts.

Tony: Can you interpret Revelation 6 where it says that stars will fall from the sky to earth as figs shaken from a tree by a strong wind?

ChatGPT

The passage you are referring to is from Revelation 6:13-14, which reads:

"And the stars of the sky fell to the earth, as figs drop from a fig tree when shaken by a strong wind. The heavens receded like a scroll being rolled up, and every mountain and island was removed from its place."

This passage is part of a larger section in the book of Revelation that describes a series of visions and events that will occur during the end times or the "day of the Lord". The language used in this passage is highly symbolic and apocalyptic, and its meaning can be difficult to discern.

The image of stars falling from the sky like figs shaken from a tree is a metaphorical description of a catastrophic event. The falling stars may represent celestial bodies or powerful entities that are being cast down or defeated. The image of the heavens receding like a scroll being rolled up may suggest a collapse of the natural order or a profound transformation of the universe.

Some interpretations of this passage suggest that it may be describing a literal astronomical event, such as a meteor shower or asteroid impact, that will have catastrophic consequences for the earth. However, it's important to remember that the book of Revelation is primarily a work of religious literature, and its primary focus is on the spiritual and symbolic meanings of the events it describes.

Overall, the passage you are referring to is a complex and highly symbolic description of a catastrophic event that will occur during the end times. Its precise meaning and interpretation may vary depending on individual perspectives and cultural contexts.

Tony: Is biblical inerrancy correct?

ChatGPT

Biblical inerrancy is a doctrine held by some Christians that the Bible is without error or fault in all its teachings, including historical, scientific, and theological matters. The concept of inerrancy is often associated with conservative or fundamentalist interpretations of the Bible.

However, the idea of biblical inerrancy is not universally accepted among Christians, and there is significant debate and disagreement over its accuracy and applicability. Some scholars and theologians argue that the Bible contains inconsistencies, errors, and contradictions, and that it should be understood as a human product that reflects the cultural and historical context in which it was written.

Other Christians hold to a more nuanced view of the Bible, recognizing that it is a complex and multifaceted text that can be interpreted in a variety of ways. These individuals may acknowledge that the Bible contains errors or inconsistencies, but still view it as a valuable and authoritative source of religious and spiritual guidance.

Ultimately, the question of whether biblical inerrancy is correct is a matter of interpretation and belief, and opinions on this issue may vary depending on individual perspectives and cultural contexts.

So, that ends my conversation with ChatGPT on this subject. We see what St. Augustine had to work with in terms of his knowledge of the universe. For him outside of space and time meant heaven in the three-tiered model of the universe depicted above, which is laughable by today's standards.

The people of St. Augustine's day didn't even know how far our sun is from earth. So, for him to say outside of space and time is to say the being (God) just comes from some kind of "beyond" to enter the physical plane.

I specifically asked ChatGPT about Revelation 6 because Neil deGrasse Tyson (a physicist) has said that the author(s) of Revelation 6 didn't know what stars were for him (them) to think they can fall from the sky to the earth as figs from a tree shaken by a strong wind.

I am not going to let ChatGPT off the hook by saying it could be a catastrophic event such as a meteorite shower. Meteorites would be bad for sure, but they wouldn't be as bad as actual stars hitting the earth. Many stars are like our sun, and if one of those hit the earth, that would literally be the end of the earth. Some of the stars out there are planets like Neptune or whatever. If one of those hits the earth, it would still be the end of the earth, but some lump of it might survive. A mere meteor is not what the author(s) of Revelation 6 intended. They wrote that the stars in the sky would fall to earth, like a bunch of little lights like glitter dust might be all over the ground if that happened.

Thus, the biblical inerrancy doctrine is shown to be false.

God Can Only Be a Lamina

We will discuss next what it means to be outside of space and time from our modern perspective.

Our universe IS space and time. For God to exist outside of space and time would require Him to be outside of our universe. What is outside of our universe? If you say more space and time, then God must be IN our universe.

So, if our universe is the ONLY space and time, then God must exist on the boundary of our universe. Let us posit that the boundary is like a mathematical boundary with zero depth but having area. In this scenario, God is a lamina.

For those who do not know what a lamina is, I asked Google for an explanation:

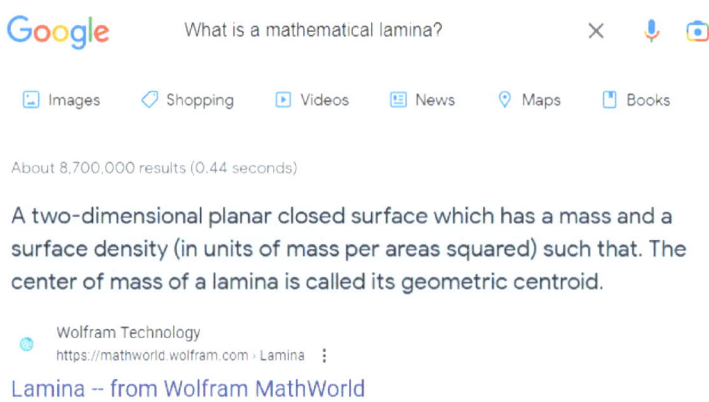

Volume is length times width times height. If height is zero, then the volume is zero. Thus, a two-dimensional surface has zero volume, and a zero volume vessel holds no mass. So, the Abrahamic God could be thought of as this lamina.

I found this image of a lamina on a sphere, which is my idea of what God would be attached to the boundary of our universe:

📖 Chapter 6.6, Problem 319E

A lamina has the shape of a portion of sphere $x^2 + y^2 + z^2 = a^2$ that lies within cone $z = \sqrt{x^2 + y^2}$. Let S be the spherical shell centered at the origin with radius a, and let C be the right circular cone with a vertex at the origin and an axis of symmetry that coincides with the z-axis. Determine the mass of the lamina if $\delta(x, y, z) = x^2 y^2 z$.

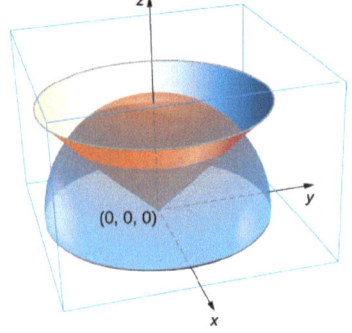

Image Credit:
https://www.bartleby.com/solution-answer/chapter-66-problem-319e-calculus-volume-3-16th-edition/9781938168079/a-lamina-has-the-shape-of-a-portion-of-sphere-x2y2z2a2-that-lies-within-cone-zx2y2-let-s-be/aa18e95e-2838-11e9-8385-02ee952b546e

This question is answered on the site given above, but I am not really interested in the solution. I just wanted you to see that the orange shaded area is on the boundary of the sphere. There is zero height on the lamina

27

everywhere it exists, so it would have zero mass. In the problem statement, a density function is given, which would allow for mass; however, this is against the belief of religion for such a lamina for God. Thus, the density function is zero for this lamina.

I have never read or heard anyone say God is a lamina before, so this is a new (and false) possibility for God's existence. Then, to be outside our universe must be to not exist since space and time only exist IN our universe.

The only other option for God to exist is in another universe with mass. If He is incorporeal, then He is a lamina on that universe's boundary since He isn't IN that universe, either, as a being with mass. To be without mass must mean He is a lamina in that other universe.

Now, either as a lamina or not existing in the nothingness of being outside our universe, God has zero power. God as an incorporeal being existing as a lamina on another universe equally rules out having any power.

The mystical or spiritual realm does not exist. We may THINK it exists because we imagine such a place in our thoughts when daydreaming or other such kinds of reverie. But, when we look for heaven using science or technology, we do not find any such spiritual realms or mystical realms in our universe. Thus, God being a lamina is as close to a solution that I can find for describing God as incorporeal. But, even then, such a God has zero power.

St. Augustine got around the question of what was God doing before He created the universe by saying time itself was God's creation so that there was no before. Similarly, where was God before creation must be answered in a like fashion–space itself was part of God's creation, so there was no space for God to exist. Yet, He is contended to exist outside of space and time, so God is a something in nothingness? Then, nothingness itself doesn't exist because God is in the nothingness.

There are other atheists who argue that God, as an infinitely powerful being, is a more complex solution to our lack of understanding of the universe. As such, we may use the razor of Occam:

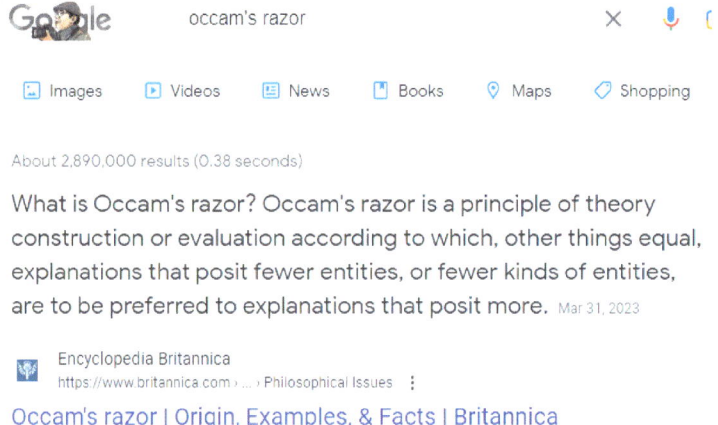

This allows us to remove God as a consideration because He is more complex than the original problem of the origin of just the universe. This is because an even more powerful God is required to create the God that created the universe. Then, who or what created that God? It's an infinitely upward spiral of more and more infinitely powerful Gods. Such a conception necessarily must be false, and the razor of Occam lops it off for us.

So, we see that however we try to choose God, we are left with a God of zero power.

Why? Well, that brings us to The God Power Equation.

The God Power Equation

$$P = \frac{mc}{t}\sqrt{v^2 + c^2}$$

That's The God Power Equation right there. Let me show you how I got it. At first, I used $E = mc^2$ along with $P=E/t$ to get the equation

$P = mc^2/t$. But, I got criticism for this because this equation didn't ALWAYS work in the universe. So, I didn't know what to do for a bit.

Well, it happened that one day I was watching a YouTube video on Einstein, and I was shown the full equation that Einstein had:

$$E^2 = (mc^2)^2 + (pc)^2$$

I didn't know what to do with momentum at first. Then, I hit upon an idea to use $p = mv$ to change the equation into one involving mass and velocity instead of momentum. When I made the substitution as before with $P=E/t$, I then got The God Power Equation as shown above.

I went back to ChatGPT to discuss with it these findings and to see if I could get it to admit that God has zero power. I put this conversation in my area on medium.com, but it gets very few views.

A Proof with ChatGPT that God Does Not Exist (Only Read If You Can Stomach the Death of God)

Preamble: I had a session with ChatGPT regarding the God Power Equation (see previous in this chapter). This equation relates mass to power, but it also involves velocity and time. I made an error in typing my version, and ChatGPT made an error in thinking P and p were the same variable (P is power and p is momentum). I deleted the chat sections involving these errors. What remains is correct analysis regarding God's power while being incorporeal.

Tony: Solve $E^2=(mc^2)^2+(pc)^2$, $P=E/t$, and $p=mv$ for P.

ChatGPT: Starting with the equation:

$$E^2 = (mc^2)^2 + (pc)^2$$

We can solve for pc by subtracting $(mc^2)^2$ from both sides:

$$(pc)^2 = E^2 - (mc^2)^2$$

Taking the square root of both sides:

$$pc = \sqrt{E^2 - (mc^2)^2}$$

Substituting p=mv:

$$pc = \sqrt{E^2 - (mc^2)^2} = \sqrt{(mc^2)^2 + (mv)^2 - (mc^2)^2} = \sqrt{(mv)^2}$$

Therefore:

$$pc = |p|v$$

where $|p|$ is the magnitude of momentum p.

Next, we can substitute pc into P=E/t:

$$P = E/t = \left(\sqrt{(mc^2)^2 + (pc)^2}\right)/t = \left(\sqrt{(mc^2)^2 + (|p|v)^2}\right)/t$$

Finally, substituting p=mv:

P = (sqrt((mc²)² + (|p|v)²))/t = (sqrt((mc²)² + (m²v²v²)))/t = (sqrt(m²c⁴))/t * sqrt(1 + (v²/c²)) = (mc²/t) * sqrt(1 + (v²/c²))

Therefore, the final solution for P is:

P = (mc²/t) * sqrt(1 + (v²/c²))

where v is the velocity of the particle.

Tony: Use your equation for power with a mass of 1 kg and a velocity of 10 m/sec. What power do you get?

ChatGPT: Using the expression P = (mc²/t) * sqrt(1 + (v²/c²)), we can plug in the given values of m = 1 kg, v = 10 m/sec, and c = 299792458 m/sec (speed of light) to find the power P:

P = (mc²/t) * sqrt(1 + (v²/c²)) P = (1 kg * (299792458 m/sec)² / 1 sec) * sqrt(1 + (10 m/sec)² / (299792458 m/sec)²) P ≈ 8.99 × 10¹⁶ watts

Therefore, the power produced by an object with mass 1 kg moving at a velocity of 10 m/sec is approximately 8.99 × 10¹⁶ watts. This is an enormous amount of power, which illustrates the enormous energy associated with high velocities close to the speed of light.

Yes, the equation $P = (mc/t)*\sqrt{v^2+c^2}$ does fix the units problem, and it does produce an answer in watts.

We can use the same values of m, v, and c as before to find the power P:

$P = (mc/t)*\sqrt{v^2+c^2}$ $P = (1 \text{ kg} * (299792458 \text{ m/sec})^2 / 1 \text{ sec}) * \sqrt{(10 \text{ m/sec})^2 + (299792458 \text{ m/sec})^2}$ $P \approx 8.99 \times 10^{16}$ watts

As before, the result is an enormous amount of power, approximately 8.99×10^{16} watts.

Tony: So, my corrected typo does produce a valid equation?

Yes, your corrected equation $P = (mc/t)*\sqrt{v^2+c^2}$ is valid, and it does produce an answer in watts.

The first term (mc/t) has units of kg m²/s³, while the second term $\sqrt{v^2+c^2}$ has units of m/s. Therefore, the product of these two terms has the units of power (watts), which have the units of kg m²/s³.

When we plug in the values of m, v, and c into the corrected equation, we get the same result as before, approximately 8.99×10^{16} watts.

Tony: Cool. So, I am not going insane.

ChatGPT: No need to worry, we all make typos from time to time. I'm glad we were able to clarify the situation and confirm that your corrected equation is valid. If you have any more questions or if there's anything else I can help you with, feel free to ask.

Tony: We both developed equations above that look different but agree. Correct?

ChatGPT: Yes, that's correct. The equations we developed above are different in their mathematical form, but they both represent the same physical concept and give the same result when evaluated with the same values of the variables. Specifically, both equations give the power P of an object with mass m moving at a velocity v, given the speed of light c, and the time t.

My equation is P = (mc²/t) * sqrt(1 + (v²/c²)), while the corrected version of your equation is P = (mc/t) * sqrt(v² + c²). Both equations involve the energy and momentum of the object, and use the speed of light c as a conversion factor between the two.

Despite the differences in their mathematical form, both equations give the same result when evaluated with the same values of m, v, c, and t.

Tony: Ok. With either version, then, may we explore the consequences of this equation?

ChatGPT: Sure, we can explore the consequences of this equation. What specific aspect of the equation would you like to investigate?

Tony: It involves the Abrahamic religions who believe that God is incorporeal. In scientific terms, what does incorporeal mean when applied to mass?

ChatGPT: In scientific terms, the word "incorporeal" is typically used to describe things that do not have a physical body or substance.

When applied to mass, the concept of incorporeality does not make sense, because mass is a physical property that is inherent to objects that have substance and occupy space. In other words, all objects that have mass are physical by definition.

Therefore, in scientific terms, it is not possible for a being or entity to have mass and be incorporeal at the same time. This means that the concept of God as an incorporeal being is not a scientific one, but rather a theological or philosophical one that lies outside the scope of scientific inquiry.

Tony: The equations we derived might now be able to say something about that. If God has zero mass, could He have infinite power in any way possible using this equation?

ChatGPT: If we assume that God has zero mass, then according to the equations we derived earlier, His power would be infinite under certain conditions.

Using my equation $P = (mc^2/t) * \sqrt{1 + (v^2/c^2)}$, we can set m to zero to get:

$P = (0 * c^2/t) * \sqrt{1 + (v^2/c^2)}$ $P = 0$

This means that if God has zero mass, His power would be zero regardless of His velocity.

However, using the corrected version of your equation $P = (mc/t) * \sqrt{v^2 + c^2}$, we can set m to zero to get:

$$P = (0 * c/t) * \sqrt{v^2 + c^2} \quad P = 0$$

Again, this means that if God has zero mass, His power would be zero regardless of His velocity.

So, based on these equations, if God has zero mass, His power would be zero. It's worth noting, however, that the question of God's existence and properties lies outside the scope of science, and is a matter of theology and philosophy.

Tony: Thank you for your analysis.

ChatGPT: You're welcome! If you have any more questions or if there's anything else I can help you with, feel free to ask.

****

Well, that ended our session. ChatGPT solved the system I solved to get a different looking equation than mine, but they actually are the same equation. With both of these equations, ChatGPT could find no way for God to have infinite power when He has zero mass. Thus, science has now shown with a legitimate equation from science based on Einstein's relativity, a power equation, and a momentum equation, that the belief that an incorporeal God (meaning zero mass) has infinite power. In fact, science shows that an incorporeal God necessarily has zero power. But, then, even faced with this, ChatGPT still says that this is outside the scope of science and belongs to theology and philosophy. Why does ChatGPT do this?

ChatGPT does this because it has been programmed to not debunk religion. It has no problem debunking the Flat Earth and other such nonsense. But, for some reason, the programmers of ChatGPT will not let it debunk religion. However, as you can see, I got it to admit that an incorporeal God has zero power, not infinite power. And, a god of zero power is a god that does not exist. Amen. I mean Q.E.D. (Q.E.D. stands for some Latin words which mathematicians use at the end of a proof to say the proof is complete.)

A Faulty Attempt to Prove God's Existence

The following YouTube video attempts to explain the first cause.

https://youtu.be/tQ2qUEQH_vs

The argument is who created us? God. Who created God? God 2. Who created God 2? God 3. Etc. With no end to this, there must be a first cause. Thus, the argument goes, that God is eternal–he always was and is unchanging and always will be.

This video has hundreds of thousands of views, and it shows that a little bit of sophistication is all that is required to dupe the masses. This is unfortunate.

A Corrected Argument Shows God Does Not Exist And Shows Science Lighting the Way

Let's show where this argument fails.

Who created us? Your parents through biological processes. Who created them? Their parents. Who created them? Their parents.

Was there a time when there weren't parents? When the species began. When was that? Quite some time ago–more than 6000 years, for sure.

Was the first set of parents Adam and Eve? No, there were multiple Adams and Eves, according to this article from 2000:

https://www.nytimes.com/2000/05/02/science/the-human-family-tree-10-adams-and-18-eves.html

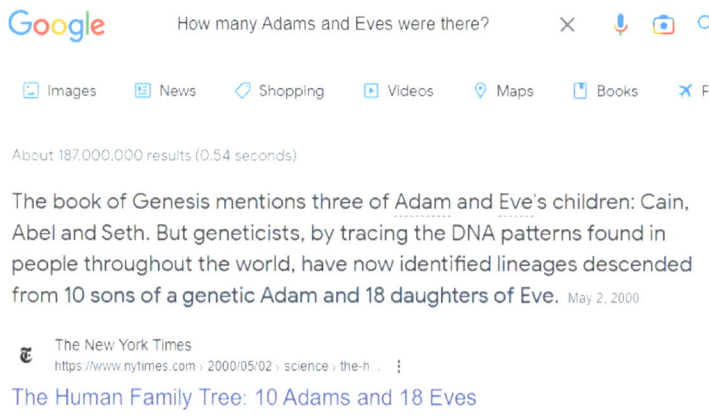

Here's another article from 2013:

https://www.livescience.com/38613-genetic-adam-and-eve-uncovered.html

Here's a relevant quote from this article:

"Almost every man alive can trace his origins to one man who lived about 135,000 years ago, new research suggests. And that ancient man likely shared the planet with the mother of all women.

The findings, detailed today (Aug. 1) in the journal Science, come from the most complete analysis of the male sex chromosome, or the Y chromosome, to date. The results overturn earlier research, which suggested that men's most recent common ancestor lived just 50,000 to 60,000 years ago.

Despite their overlap in time, ancient "Adam" and ancient "Eve" probably didn't even live near each other, let alone mate. [The 10 Biggest Mysteries of the First Humans]"

Is the universe eternal? No, there is strong evidence for the Big Bang, so the universe had a beginning about 13.8 billion years ago.

God is eternal, right? No, God doesn't exist. If He exists outside of space and time, there's no way to prove that. But, He cannot exist in our universe as an incorporeal being because the God Power Equation shows that zero mass equates to zero power. This equation holds sway ***in our universe***.

What if I don't believe that? It isn't necessary for you to believe in the God Power Equation or not. It is correct because it is based on Einstein's relativity and two definitions. To prove the God Power Equation wrong, you'd have to overturn relativity or either of the two definitions. If you do any of these things, you'd turn physics on its head. (Note: You aren't going to do any of this–you are simply incorrect about your belief.)

What follows is a series of articles I wrote on the internet related to The God Power Equation. I make no attempt to edit them into a coherent whole, but the articles taken together do constitute a coherent whole.

By Raphael — Own work, J1m1mayers, 1 January 1518, CC BY-SA 4.0, https://commons.wikimedia.org/w/index.php?curid=63865341

Deriving The God Power Equation

We Explore the Incorporeal God Hypothesis

We will start with the full mass-energy-momentum equation:

$$E^2 = (mc^2)^2 + (pc)^2$$

https://www.britannica.com/video/185388/equation-theory-energy-relativity-mc

So, this is Einstein's full equation (his $E=mc^2$ is just a partial equation). It works in all conditions in the universe regardless of how fast particles are moving in reference to each other. You can watch Britannica's video with the link below the image.

Another equation we will need is the very simple $P = E/t$ or power is energy divided by time.

So, we replace E in the first equation with Pt from the second equation. This gives us $(Pt)^2 = (mc^2)^2 + (pc)^2$. Next, we divide both sides by t^2 to get the next equation:

$$P^2 = (mc^2)^2/t^2 + (pc)^2/t^2$$

Now, if mass is zero, which is the case for incorporeal spirits and gods and the like, we can eliminate that term:

$$P^2 = (pc)^2/t^2$$

Note that capital P is for power and lowercase p is for momentum. Next, we have an equation for momentum!! It is $p = mv$, which is momentum is mass times velocity. We replace the lowercase p in that last equation, then, as follows:

$$P^2 = (mvc)^2/t^2$$

Finally, if m is zero, which it is for incorporeal beings, this term also eliminates.

$$P^2 = 0$$

We take the square root of both sides to get the final God power equation:

P = 0

Thus, the almighty god has a zero power level if He has no mass.

This is now an irrefutable equation because it uses all the valid equations of physics. Thus, it is now proven that God's power or the power of any incorporeal being is zero in our universe.

We now show through a Google screenshot who claims that God is incorporeal:

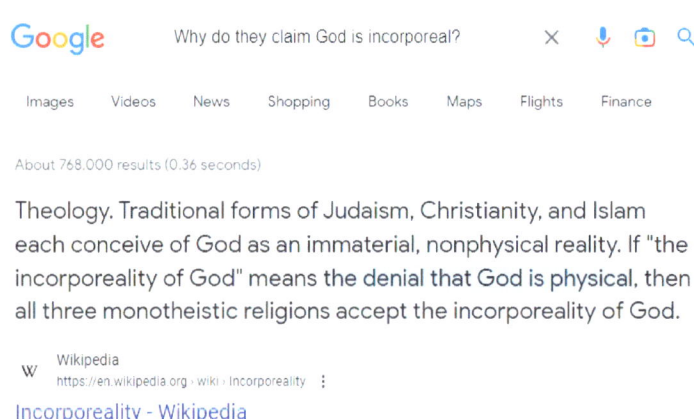

https://en.wikipedia.org/wiki/Incorporeality

$$E^2 = (mc^2)^2 + (pc)^2 \quad (1)$$

$$P = \frac{E}{t} \quad (2)$$

$$p = mv \quad (3)$$

Solve the system of 3 eq's for P and factor.

$$E^2 = (mc^2)^2 + (pc)^2$$
$$(Pt)^2 = (mc^2)^2 + (mvc)^2$$
$$P^2 t^2 = m^2 c^4 + m^2 v^2 c^2$$
$$P^2 = \frac{m^2 c^2}{t^2}[c^2 + v^2]$$
$$P = \frac{mc}{t}\sqrt{c^2 + v^2} \quad \leftarrow \text{God's Power Equation}$$

Here, in the above photo, I show the full derivation of God's Power Equation. Technically, the last step should show a +- symbol, but I eliminated it from display because negative power could only come from negative mass, which is either gravity (according to Michio Kaku) or it's antimatter mass from a different universe. As I am attempting to ascertain God's power in our own universe, the negative answer is not relevant at best. But, I mention it here for the sake of completeness.

Now, it is claimed that God's power is infinite. We see from this equation that this is possible in three ways: infinite mass, zero time, or infinite velocity.

God's mass is zero as claimed by the religions, so that rules out this way. Zero time is impossible as even a quantum fluctuation doesn't take zero time. And finally, infinite velocity is impossible as God would blow past our universe before he even noticed someone had a prayer. Also, nothing travels faster than the speed of light, and this includes God. Thus, this equation shows God's power is not infinite. The only logical answer provided by this equation is that since God's mass is zero, His power is zero. This means God does not exist.

Thanks to Albert Einstein for providing the bedrock equation in this God Power equation. But, the derivation is mine to the final equation. And, I have displayed it in my own handwriting.

Update: I just tried GPT-4, and asked it to solve the system above. It couldn't do it!!

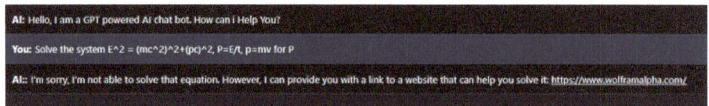

I guess I am smarter than GPT-4 lol.

If God's Power Isn't Infinite, What Is His Maximum Power???

Science Can NOW Answer This!!!

We will start with The God Power Equation, derived by Tony Berard in 2023. This equation will allow us to calculate God's power under a variety of circumstances. We seek God. No, just kidding. We seek to know what the maximum power of God could be.

Of course, the faithful CLAIM that God's power is infinite. But, let's test that claim, shall we?

$$P = \frac{mc}{t}\sqrt{v^2 + c^2}$$

The God Power Equation

So, we would like to know the mass of the universe as this is the greatest possible mass that we can plug into The God Power Equation:

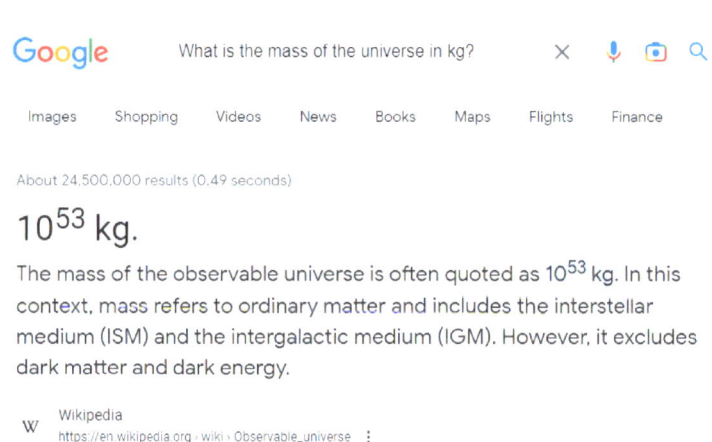

So, this is the number we will use. We won't speculate on how much mass there is in the dark matter and dark energy because scientists are still working on figuring this out.

Next, we need to know what the smallest amount of time possible for God to act. Google is good at finding out such trivialities. So, I asked Google for this:

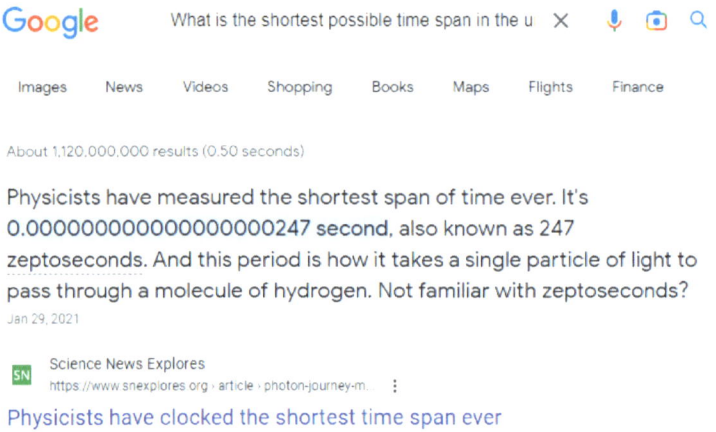

Next, we have that c sitting there in the equation. We need to know its value in meters per second:

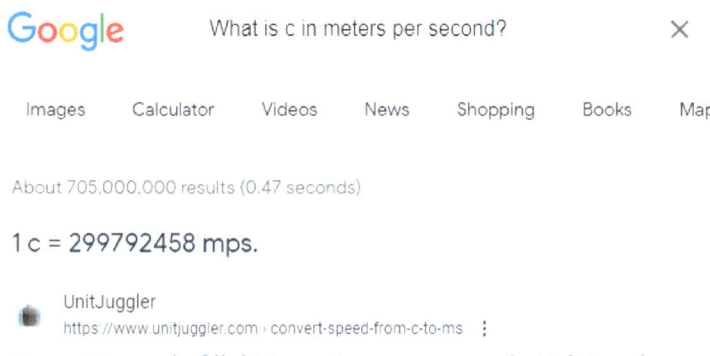

Now, we are told that nothing travels faster than the speed of light. Space can expand faster than the speed of light, but God isn't space itself. If God exists, He needs to be inside our universe to do anything. And, He needs to be close enough to us to answer our prayers even allowing for the limited speed of light. He also needs to be composed of matter such that He can interact with it like we do to do things for us when answering our prayers. Thus, we declare that God is bound by the laws of our universe, and He cannot exceed the speed of light, c.

There, we have identified all of the variables in our equation. We simply plug it all in and simplify. That's great — we simply simplify. Note: 1 second = 10^{21} zeptoseconds.

Pmax = (mass of universe x c / smallest unit of time) x sqrt($c^2 + c^2$) = sqrt(2) x mass of universe x c^2 / smallest unit of time = 1.414 x 10^{53} kg x (299792458 m/s)² / 247 zeptoseconds x (10^{21} zeptoseconds / 1 second) = 5.1458775830877030059935916169288 x 10^{88} Watts

So, this is about five hundred billionths of a Googol Watt. This is the amount of power that would be generated if the entire mass of the universe were converted to energy and dispensed in 247 zeptoseconds. One can clearly see that that is a lot of power. But, one can also clearly see that it isn't an infinite amount of power as religions like to CLAIM. Claims can be tested and found to be false.

Another claim of religion is that God (aka Yahweh, Allah, Elohim, and Shiva and Vishnu) are incorporeal, which means that they have no mass. According to The God Power Equation, such a being has zero power. A zero power God is a god that does not exist.

Thus, the website quoted at the top that declares that God is infinite in power and is able is found to be incorrect. Thanks for playing.

Next, we have another article I wrote:

Who Is the Supreme Deity: Yahweh or Shiva?

Science Can Now Answer This!!

Yahweh   Shiva

Image Credits:

Yahweh: By Rama, CC BY-SA 3.0 fr, https://commons.wikimedia.org/w/index.php?curid=87617466 Shiva: By Thejas Panarkandy from India   Murudeshwara Statue, CC BY-SA

2.0, https://commons.wikimedia.org/w/index.php?curid=9761046

You've all heard of Einstein's famous E=mc² equation. If we use this to find God's power, we combine it with P = E/t to get P = mc²/t. The incorporeal god means in scientific terms that such a non-physical being has zero mass. Plug zero in for *m* in this power equation, and you'll see that God has zero power because zero times anything is zero.

I received criticism of this in earlier articles because *E=mc²* doesn't ALWAYS work. But, Einstein had a fuller version of his equation that DOES WORK in all situations in our universe. It is the following:

$E^2 = (mc^2)^2 + (pc)^2$

This is the full relativity equation, and that lowercase *p* is momentum. So, we have two substitutions to make: one with the equation *P = E/t* and the other with *p = mv*. That second equation is momentum equals mass times velocity.

When we make the substitutions, we get the following:

$$P = \frac{mc}{t}\sqrt{v^2 + c^2}$$

The God Power Equation

Let us say that v is zero. That makes the c^2 the only thing left under the radical. The square root of the square of c is just c. We multiply that by the c standing next to the m in the numerator over there, and what do we get? We get $P = mc^2/t$, which is what the other equation said. Thus, we can now account for velocity in our God Power Equation.

This equation is not complex. It can be understood by any student who has completed their first algebra course in high school or college. It only has squares and square roots, which are rather low level objects in mathematics.

So, we have here an equation that debunks God, and it is a very simple equation that can be understood by hordes of people, not just the Ph.D.'s making advancements in mathematics and physics. This is a big deal!!

All the Abrahamic Religions claim that God is incorporeal. Thus, the God in seven religions has no mass: Christianity, Islam, Judaism, and four others. This scientific equation that is easily understood shows that the incorporeal God has no power.

Now, let's show who promotes the incorporeal god.

The Christian Library:

God Is Incorporeal, Immutable, Invisible, Glorious School of Theology Series: Lecture 8

God is incorporeal. That means to say that he has no body-a reference here to the fact that the Bible often speaks of...

https://www.christianstudylibrary.org/article/god-incorporeal-immutable-invisible-glorious

Dr. William Lane Craig:

God's Incorporeality | Reasonable Faith

KEVIN HARRIS: Bill, you are making some great progress on your systematic philosophical theology. You are writing that…

https://www.reasonablefaith.org/media/reasonable-faith-podcast/gods-incorporeality

Kermit Zarley:

Is God Incorporeal or Not?

The word "corporeal" is usually understood as referring to the physical body of a human being, especially as…

https://www.patheos.com/blogs/kermitzarleyblog/2015/09/is-god-incorporeal-or-not/

From Kermit Zarley's site:

Thus, orthodox Christian theology asserts that God is incorporeal, meaning he does not consist of physical. Before we go any further, we must answer the question, Who is God? I believe the God of the Bible, the God of Abraham, Isaac, and Jacob, is YHWH (perhaps pronounced and spelled "Yahweh"), and this God Jesus calls "Father." According to the Bible and Jesus, Yahweh, the Father, is the one true God (e.g., Duet. 6.4–6; John 17.3).

RE: Online

Incorporeality / Immateriality of God - RE:ONLINE

The idea of the incorporeality of God is present in the Abrahamic religions. It means that God has no physical body or…

https://www.reonline.org.uk/teaching-resources/16-plus-philosophy/incorporeality-immateriality-of-god/

I didn't know who RE: Online was, so I found this quote on their site:

Here's a few more for good measure:

These places derive money and support from the public who BELIEVE them. I am trying to show that this is misguided. An incorporeal god is a god with no power. This scientific equation shows it very clearly and easily.

Here's a book on Amazon about the incorporeal god:

The Incorporeal God: An Insight Into The Higher Realms
https://www.amazon.com/Incorporeal-God-Insight-Higher-Realms-ebook/dp/B07L3NZLC9

So, the above author claims to take the reader to higher realms. Higher realms of what? More incorporealness? Please, let science now silence such people who promote this nonsense.

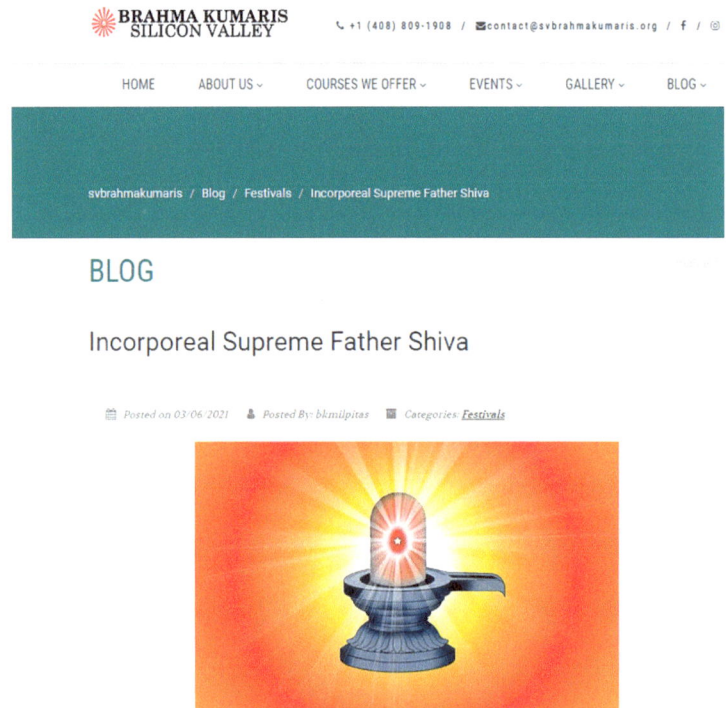

https://svbrahmakumaris.org/incorporeal-supreme-father-shiva/

I don't know who Shiva is, so I thought I'd ask Google:

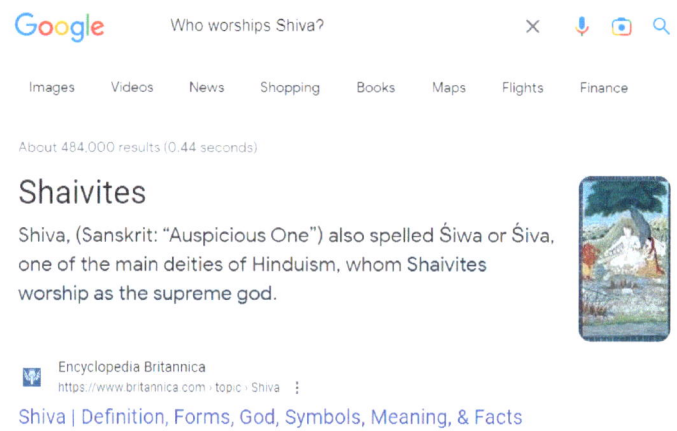

https://www.britannica.com/topic/Shiva

So, I clicked on the Britannica article, and I got this image on the next page with an explanation.

So, this equation has disproved two supreme incorporeal gods: Yahweh and Shiva. Both have zero power. Thus, it is a tie — they both have no power.

Shiva

The god Shiva in the garb of a mendicant, South Indian bronze from Tiruvengadu, Tamil Nadu, early 11th century; in the Thanjavur Museum and Art Gallery, Tamil Nadu.

P. Chandra

SHARE

Finally, here's a site with specious claims:

Face of God? Archaeologist claims to find 10th cent. BCE graven images of Yahweh
https://www.timesofisrael.com/face-of-god-archaeologist-claims-to-find-10th-cent-bce-graven-images-of-yahweh/

Next, here is another article I wrote on medium.com.

Science Can Now Prove God Has No Power

(Me and Einstein and a Little Algebra)

Einstein with his second wife, Elsa, in 1921

$$P = \frac{mc}{t}\sqrt{v^2 + c^2}$$

The above equation is what I call the God Power Equation. I wrote about it here:

https://medium.com/the-globally-responsible-atheist/the-god-power-equation-a115fbbe528d

P is power, p is momentum, m is mass, c is the speed of light, v is velocity, and t is time. We get this equation by solving the following system of equations from physics:

$$\text{Solve } E^2 = (mc^2)^2 + (pc)^2, \ P = E/t, \text{ and } p = mv \text{ for } P.$$

The first equation is due to Einstein. The second equation is the definition of power in terms of energy per unit time. The third equation states that momentum is mass times velocity. Solving this system of three equations requires nothing more than beginning algebra — possibly a little bit of intermediate algebra because of the presence of the squaring exponents.

I wrote a paper a while ago that used Einstein's E = mc² formula, but I got criticism because that equation did not work in every situation. Einstein's other equation with E² in it above is the full equation and does work in all situations in the universe.

Because of this fuller equation having momentum in it, we need to expand the system to the three equations depicted above and to solve it for P (power).

I did this and got the God Power Equation. So, armed with this, I asked ChatGPT about it.

I observed that its calculational ability had been significantly increased. So, I had a full conversation with it in this article:

A Proof with ChatGPT that God Does Not Exist (Only Read If You Can Stomach the Death of God)

https://tonyberard.medium.com/a-proof-with-chatgpt-that-god-does-not-exist-only-read-if-you-can-stomach-the-death-of-god-418962b73766

The key part of this conversation is the following:

$P = (0 * c/t) * sqrt(v^2 + c^2)$ $P = 0$

Again, this means that if God has zero mass, His power would be zero regardless of His velocity.

So, based on these equations, if God has zero mass, His power would be zero. It's worth noting, however, that the question of God's existence and properties lies outside the scope of science, and is a matter of theology and philosophy.

I gave ChatGPT the opportunity to find a way in light of this true equation from physics (and it recognized it as the governing equation) to find a way for a god of zero mass (i.e. an incorporeal god) to have infinite power. It could only conclude that a god of zero mass has zero power. This is the correct conclusion based on this equation from physics that describes how power, mass, time, and velocity are related in our universe.

The programmers of ChatGPT tell it to always say questions of God and religion and miracles and such lie outside the scope of science. But, no they don't lie outside the scope of science — not any more.

We needed Einstein to come up with the E^2 equation above from relativity. We couldn't prove an incorporeal god has zero power without it. But, we needed me as well to doggedly pursue this path until I finally found the God Power Equation which emphatically proves an incorporeal being has zero power regardless of his velocity as ChatGPT has said.

The standard things I hear are that you can't

prove a negative (which is false because you can), and that God lies outside the scope of science. Now, God falls within the scope of science with this God Power Equation. It has shown clearly and convincingly that zero mass means zero power. So, we have in the first place a contradiction in that God does not have infinite power because of this equation. This casts doubt on the belief in God. But, there is more. Zero mass actually means zero power. So, not only does God NOT have infinite power, he actually has zero power. A god with zero power is a god that does not exist.

So, science now fully debunks the incorporeal god belief with infinite power. Yahweh started out as a storm god. He had a mother and father (Asherah and El). He had 69 brothers, including Ba'al. Over time, El, Asherah, Ba'al and the rest were denied to exist leaving only Yahweh. So, Yahweh went from being a living, breathing being to becoming incorporeal (no mass) to stave off criticisms to attempt to become "unfalsifiable."

But, now, science has this relationship between mass, power, velocity, and time understood in the manifestation of the God Power Equation. And, this relationship shows that zero mass equates to zero power. Nietzsche was correct in that God is dead. He really never existed in the first place, but God dying is sufficient for the purposes of the Abrahamic religions, among them Christianity.

Debunking the Trinity

The transitive law in math says if A=B and B=C, then A=C. In plain words, if two things each equal a third thing, then they must equal each other. This is due to what "equals" means. Equals means the same as. So, we come to The Trinity. In the following diagram, we have three things each equaling a fourth thing. Yet, these three things somehow do not equal each other. Really? How is this so?

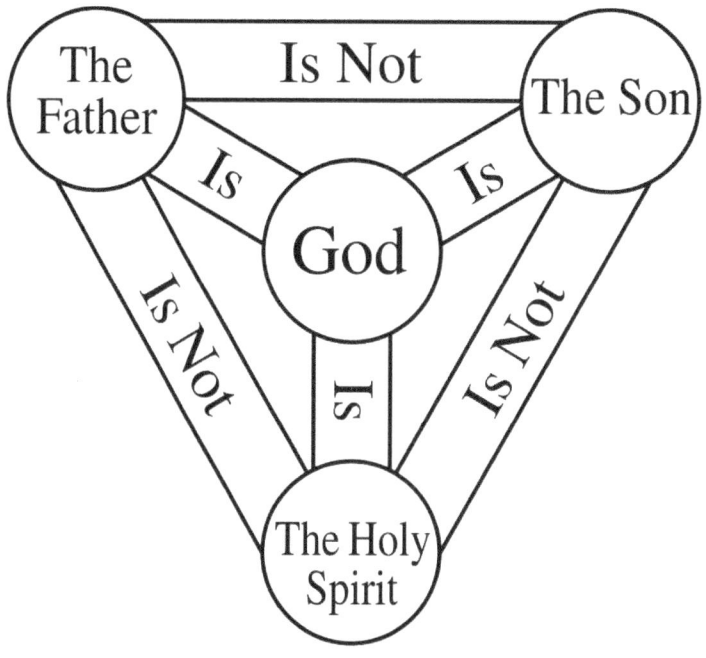

The only way to wiggle out of this is to allow God (the thing that the others is equal to) to change into something equal to one of the other three when called for. But, this is impossible because of the immutability (or constancy) of God. Thus, God does not change. So, The Father = God, and The Son = God, and The Holy Spirit = God (and these are all equal to the very same unchanging God), but they do not equal each other is just nonsense. Thus, the Trinity is debunked because it violates the transitive law in mathematics (something taught in arithmetic classes before the first algebra course).

Proof That God Does Not Exist
The Proof with Full Explanations

The Almighty God

This very paper is on Medium, but I wanted to write it again with a complete explanation because so many people on here weren't correctly understanding this proof. My new comments will be in italics like this is, and all the plain font text will be the original article. The first explanation, then, is that this photo is not of God. It is an artist's interpretation of God. Here is the link for it:

https://pixabay.com/illustrations/light-ray-blaze-gradient-734436/

One of the responses I got in comments on Medium was that one couldn't prove God does not exist, either, which was why the respondent was an agnostic.

I have had numerous comments to the effect saying that you cannot prove a negative (like this person had wrongly claimed). I had thought that you could prove a negative, but these comments stating that you couldn't made me doubt myself. So, I researched it. Here is the results of my research:

Yes, You Can Prove a Negative:
https://tonyberard.medium.com/yes-you-can-prove-a-negative-9d30fbc73709

So, dear agnostic, here is a proof that God does not exist.

It must be the case that God exists or does not exist. If we prove one, we automatically disprove the other and vice versa.

Let us therefore try to prove that God exists.

God is all-powerful — nothing is more powerful than God.

I got some responses to this line of the proof that I somehow have faith that God is all-powerful. No, I don't subscribe to that. This line of the proof is here as an assumption. This means we will treat it as being true for the time being. Other statements will be used against it that create a crisis in the proof.

So, how much power does God have, then?
Infinite is the reply.

There are many infinities, so which infinity of power does God have?
The largest one, of course.

It can be shown that there is no largest infinity (no largest prime number, no largest set, no largest area, no largest volume, no largest power, etc.) The attempt to say THAT one is the largest is countered by a method to generate a larger one.

Therefore, the attribute of largest power possible is not possible to assign to God.

Therefore, the proof that God exists fails. Thus, God does not exist.

Georg Cantor was approached by the church because his work with infinity was against God. Here's what I can find on this assertion:

Why did Georg Cantor's work face so much opposition?
https://www.quora.com/Why-did-Georg-Cantors-work-face-so-much-opposition

Cantor attempted to create an absolute infinity which could not be surpassed and was equivalent to God's infinity. However, this was not possible as it is an infinity just like any other.

Absolute Infinite - Wikipedia
https://en.wikipedia.org/wiki/Absolute_Infinite

Thus, God's infinity can be bested, which shows that God does not exist from the above proof.

In Conclusion

The bible doesn't get anything right. The accounts it has of various things like the flood and creation were myths simply borrowed from other cultures.

God doesn't exist. He was invented as a means to allow a ruling class to exist. Nowadays, such a mechanism isn't necessary as we have political systems now and dictators that do the job of social control quite well. Plus, monarchs saw that popes saying that God gave them authority was very useful. Monarchs started saying God gave them authority as well.

Nevertheless, God survives in the belief systems of many people today. My contribution to this debate is to provide the equation that relates mass and power (and time and velocity) to the belief that an incorporeal God has infinite power. The God Power Equation shows very simply (and multiplication by zero really is as simple as it can get) that any incorporeal god has zero power in our universe. Zero power is not just not infinitely powerful, it means that God does not exist. I also provided proof by contradiction that God doesn't exist.

There will be believers who dismiss this fact claiming God makes all the laws including the God Power Equation and is above them. They'll claim proofs by contradiction don't work on God. But, these are false steps for the believer. God has no power in our universe because He has no mass. If you want to say God is outside of space and time, then, He can only be a lamina with zero mass and zero power. God cannot exist outside of space and time because there is no space and no time outside of space and time for God to exist in.

Thanks for reading me.

About the Author

I have a bachelor's degree in mathematics from Lawrence Technological University in Southfield, MI. I have written a number of books before, and they are available here on Amazon. One is Points, Lines, and Conic Sections: A Sequel to College Algebra.

I invent board games. My work is almost available at www.fightchess.com. One of my best games (Fighting Chess) is our first game we are doing.

I developed a superior rating system to the Elo rating system used in chess. We will use my rating system for the ratings we have on our site. Our rating system can tell you the rating of an unknown player in a single tournament. We've all heard of players who are playing "above their rating" and are waiting for "their rating to catch up with them." That is a thing of the past!!

I also developed a new kind of tournament called the Block Transition Tournament. This new type of tournament design is very flexible. It also can enable you to have the top two performers meet up in the final round (or final three rounds) to play head to head for all the marbles. This is a major advance in how tournaments are conducted.

I do have another board game I invented that's called Tines and Barbs, which is a very deep game. It has rotating pieces, guns that wound and kill, shields that block but can be removed, and even a magic wand carried by the king (called The Grand Barb). I will be a very happy man when this game becomes available on our site.

I play the piano, but I haven't done so in a number of years. One day, I might compose some music for our board gaming site.

I am a big fan of fairness and truth, which is the main reason I wrote this book. The truth of the God Power Equation really resonates with me. I am grateful for having discovered it.